Fur Seals

and Other Pinnipeds

Concept and Product Development: Editorial Options, Inc.
Series Designer: Karen Donica
Book Author: Lome Piasetsky

For information on other World Book
products, visit us at our Web site at
http://www.worldbook.com

For information on sales to schools and libraries
in the United States, call 1-800-975-3250.

For information on sales to schools and libraries
in Canada, call 1-800-837-5365.

World Book, Inc.
233 N. Michigan Ave.
Chicago, IL 60601

Library of Congress Cataloging-in-Publication Data

Piasetsky, Lome
    Fur seals and other pinnipeds / [book author, Lome Piasetsky].
      p. cm.—(World Book's animals of the world)
    Summary: Questions and answers explore the world of pinnipeds, with an emphasis on seals.
    ISBN 0-7166-1202-X -- ISBN 0-7166-1200-3 (set)
    1. Seals (Animals)—Juvenile literature. 2. Pinnipedia—Juvenile literature [I. Seals
(Animals)—Miscellanea. 2. Pinnipeds—Miscellanea. 3. Questions and answers.] I. World Book,
Inc. II. Title. III. Series.
    QL737.P6 P52 2000
    599.79—dc21   00-021633

Printed in Singapore
1 2 3 4 5 6 7 8 9 05 04 03 02 01 00

World Book's Animals of the World

Fur Seals
and Other Pinnipeds

What happens when I lose my fur coat?

World Book, Inc.
A Scott Fetzer Company
Chicago

Contents

How could I get lost in a crowd?

Who's the king of the ice pack?

How deep can I dive?

What Is a Pinniped?

The fur seal you see here is a pinniped *(PIHN uh pehd)*. The word *pinniped* means *fin-footed*. A pinniped is an animal with four "fin-feet," or flippers. Fur seals, sea lions, and elephant seals are pinnipeds. Walruses are pinnipeds, too. Flippers help all these animals be very good swimmers.

All pinnipeds are mammals. They are warm-blooded, and hair or fur covers their bodies. Their pups, or babies, feed on the milk of their mothers. Although pinnipeds spend much of the time in water, they need to breathe air, just as you do.

Fur seal

Where in the World Do Pinnipeds Live?

Pinnipeds are sea mammals. Most live where the climate is cool or cold. Pinnipeds like to stay around ocean islands and along the coasts of continents. A few kinds live in lakes. Pinnipeds eat fish and other sea creatures.

Some pinnipeds leave their home bases at times. They may migrate, or travel to warmer places, before it's time to give birth to their young. Pinnipeds may also migrate to find new supplies of food. Migrating pinnipeds sometimes travel great distances. But they almost always return to their home bases.

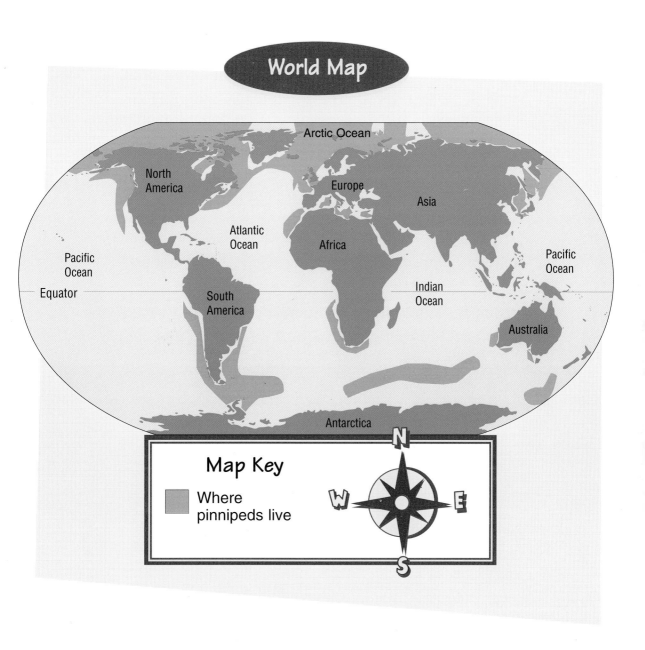

World Map

Arctic Ocean

North America

Europe

Asia

Atlantic Ocean

Africa

Pacific Ocean

Pacific Ocean

Equator

South America

Indian Ocean

Australia

Antarctica

Map Key

Where pinnipeds live

N

W E

S

9

What Does Fur Do for a Fur Seal?

A fur seal has not one, but two layers of fur that cover its body. This thick fur keeps the animal warm and dry.

Fur seals can swim very fast. They paddle with their front flippers. Fur seals stick out their heads and necks to steer. On land, fur seals use all four flippers. See how this seal is sitting. Its hind flippers are turned forward and down. When the seal walks, it uses all four flippers to do so.

Underwater, fur seals have other features that help them find food and avoid enemies. Their small ear flaps curl over their ear openings. This helps keep water out. Their large, round eyes help them see in dark water. Their nostrils close when they dive, and they hold their breath until they need to come up for air. Their whiskers probably help them find food by touch.

Northern fur seal

When Is It the Right Time for a Swim?

For a fur seal, the right time for a swim is almost always. Like all pinnipeds, fur seals spend most of their time in the water. Their body shape makes them strong swimmers and divers.

Fur seals search for their food in the water. Most of the time, they find fish and other sea animals near the surface of the water. Sometimes, they must dive to find food. Some of the dives are shallow, but others are quite deep.

Fur seals may spend days at a time in the water. They are most active during the evening, at night, and early in the morning. They sleep during the middle of the day, floating on their sides.

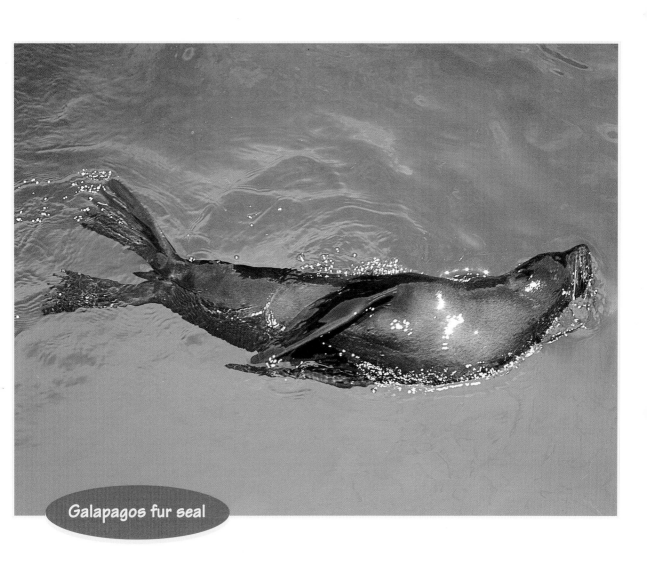

Galapagos fur seal

When Is It Time to Haul Out?

Fur seals spend a lot of time in the water. But sometimes they need to "haul out." When a fur seal hauls out, it uses its front flippers to pull itself up onto land or ice.

Fur seals haul out to rest and warm up in the sun. They may also haul out to escape from enemies, such as killer whales or sharks.

There are two other important times when fur seals leave the water. They haul out at molting time. When seals molt, they shed their old fur coats and grow new ones. And they haul out to have their pups.

Fur seals come out of the water for good reasons, but getting out may not be so easy. A male Cape fur seal, also called a South African fur seal, may be more than 7 feet (2 meters) long and weigh more than 660 pounds (300 kilograms). That's a lot of body to be dragging out of the water!

Cape fur seals

What Is Under All That Fur?

If you could see inside a fur seal's body, here is how its bones would look.

Look carefully at the fur seal's four legs. Above the ankles, the legs are inside the seal's body. The ankles and feet form large flippers. Each flipper has five toes. The toes are webbed, which means that they are connected by skin. The webbed feet act like paddles, helping the seals swim.

A flipper

The spine, or backbone, of a fur seal is special, too. A fur seal uses its spine to make snakelike motions. These motions help propel the seal through the water.

Seal Skeleton

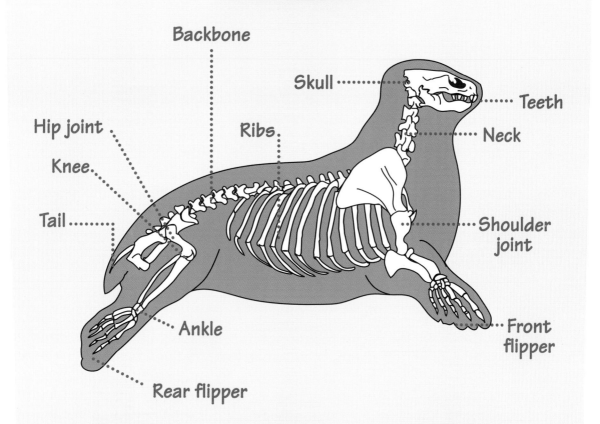

Backbone

Skull

Teeth

Hip joint

Ribs

Neck

Knee

Tail

Shoulder joint

Ankle

Front flipper

Rear flipper

What Are Fur Seal Pups Like?

Fur seal pups are born in the spring or summer. Mother seals usually give birth to only one pup at a time.

Newborn seal pups are more developed than human babies are at birth. Fur seal pups can walk and swim, but they wait a few weeks before they go into the water. Their eyes are open, and they can make sounds. Fine, soft fur covers their bodies. When the pups grow up, they will have fur like their parents have.

At first, a pup drinks only its mother's rich milk. But pups grow up fast. Very soon, they are able to find their own food in the water.

Baby Galapagos
fur seals

How Does a Mother Find Her Pup?

Fur seal pups are born on land, in a place called a rookery. Thousands of mothers and their babies may crowd into a single rookery.

A mother Cape fur seal stays with her pup for most of the first week of its life. Five to seven days after giving birth, she goes off to sea to find food for herself. She eats and eats. It takes a lot of energy to feed a growing pup!

How does the mother find her own pup when she returns to the rookery? After she hauls out, the mother calls to her pup. The pup hears her and calls back. Each mother and pup have their own special call. This helps them to find each other in the crowded rookery.

When a mother and her pup find each other, they often rub noses. Each mother and pup also have a special smell. A mother knows that she has found the right pup by its smell.

Rookery of Cape fur seals

Are Sea Lions Really Lions?

No, but sea lions do have some things in common with those big cats. That's why *lion* is part of their name. Male lions have manes. So do many kinds of male sea lions. Sometimes, male sea lions even roar like lions.

Among the pinnipeds, sea lions and fur seals are close cousins. Both swim with their front flippers and can walk on all fours. Both have ear flaps that cover their ear openings. Most sea lions are bigger than fur seals, however. The coats of sea lions are not so thick, but they do have more blubber. This layer of fat under the skin keeps the sea lions warm. It also gives them energy.

South American
sea lion

Are Sea Lions Smart?

Many people think sea lions are quite smart. The reason is that sea lions can learn to do many things. And they remember what they learn.

Sea lions can follow word signals and hand commands. They can also find objects underwater. Sea lions can jump through hoops and clap their flippers together.

Sea lions are most at home in oceans and seas. But you may have seen some in a zoo, a circus, or an aquarium. Perhaps you have seen a California sea lion balancing a ball on its nose and getting rewarded with a fish. California sea lions are good performers.

California sea lions

How Can Sea Lions Help People?

The sea lion in this picture is hard at work. It is attaching a cable to a sunken model of a torpedo. The U.S. Navy will pull up the torpedo. The navy has been using sea lions and dolphins to find and help recover objects that are sunk deep in the ocean. Since sea lions are such great divers, they can do the work better than humans.

Sea lions are also helping scientists study whales. Whales are hard to observe because they spend almost all their time underwater. Even if scientists get into the water with whales, the animals might sense the humans. The whales might respond to the people rather than behaving as they usually do.

Whales are used to being in the water with sea lions. So scientists are training these animals to wear cameras and to videotape whales. It will be interesting to see how successful sea lions are in becoming "whale watchers."

Sea lion

Which Pinniped Is the Nose-iest?

The seal you see here is an elephant seal. Its nose looks like an elephant's trunk, but it is shorter. Only male elephant seals have trunks for noses, which they use to attract female elephant seals.

Elephant seals are the biggest pinnipeds. Whales are the only sea mammals that are larger. Southern elephant seals may weigh up to 8,800 pounds (4,000 kilograms), and males may be more than 16 feet (4.9 meters) long.

Elephant seals cannot walk on all fours. They cannot rotate their hind flippers the way fur seals can. Elephant seals use their front flippers and strong stomach muscles to pull themselves forward, which makes them look like big caterpillars moving on land. Even though elephant seals cannot move very fast or far on land, they are excellent divers and swimmers.

When elephant seals molt, they lose large patches of skin and hair. Some females wallow in mud as they molt.

Northern elephant seal

How Deep Can an Elephant Seal Dive?

Elephant seals have been known to dive more than 5,000 feet (1,524 meters). That's almost a mile!

In fact, elephant seals spend most of their time at sea. They do not need to rest very long between dives. Sometimes, they dive for more than an hour without coming up. That's a long time to dive without breathing! How do they do it?

Elephant seals store a lot of oxygen in their blood. They store more oxygen than most other kinds of pinnipeds. They use this oxygen while they are underwater. This allows them to go for a long time without breathing air.

During their dives, elephant seals look for food. Their favorite foods are deepwater fish and squid. But they also eat small sharks and octopuses.

Elephant seal

Which Pinniped Is a Homebody?

Many kinds of pinnipeds migrate long distances to find food or breeding grounds. But one seal that usually stays close to home is the harbor seal.

Harbor seals live mainly along shorelines. They can be seen along both the east and west coasts of the United States. Their favorite places are harbors.

Harbor seals usually haul out on rocks, piers, and sandbanks. Females often give birth where tides go in and out, away from boats and other dangers. Their pups can swim when they are just a few minutes old. But harbor seal pups rarely swim alone. Their mothers go with them. If the adult senses danger, she may take her pup in her mouth and dive underwater with it.

A pup might even get a ride on its mother's back while she hunts for food.

Harbor seal

How Did the Ribbon Seal Get Its Name?

A male ribbon seal has dark brown fur with white or light yellow bands around its neck, its front flippers, and its lower back. These bands look like ribbons.

A large group of these seals lives in the Bering Sea, which is part of the North Pacific Ocean. Ribbon seals like to haul out on pack ice. Pack ice is a large mass of floating ice.

Ribbon seals are speedy walkers on ice. Like many other pinnipeds, they use their front flippers for walking and their back flippers for swimming. On land, ribbon seals can move faster than a human running at top speed.

Ribbon seal

Which Pinniped Wears Rings?

A ribbon seal may look as if it wears ribbons. But a ringed seal appears to wear rings. And it wears these rings all over its body!

Ringed seals are among the smallest pinnipeds. They live mostly in the Arctic, where there is a lot of ice. When ringed seals are breeding, they like fast ice. Fast ice is ice that is attached to land—rather than ice that is floating in the water. Nonbreeding adults like pack ice. Of course, there is a lot of water under the ice, and that is where ringed seals hunt for food.

Ringed seals have sharp claws on their front flippers. They use their claws to make breathing holes in the ice. When they stop hunting and come up for air, the seals must be very careful. A polar bear may be waiting at the hole for its next meal!

Ringed seal

Why Do Ringed Seals Build Ice Houses?

A mother ringed seal cares for her pup in a special way. She builds her pup an ice house!

Here you see how the mother builds the ice house. She swims under the ice until she finds a place where it has cracked. Where there are cracks, there are also holes. The mother seal comes up through a hole. She then uses her sharp claws to dig in the ice and snow. She hollows out a place to give birth and to care for her pup. The ice house is warm and snug—and safe!

Ringed seal pups are small. They do not have much blubber on their bodies. To keep warm, they cuddle close to their mothers. When they get a little older, the pups may dig tunnels to other nearby ice houses.

A Ringed Seal's Ice House

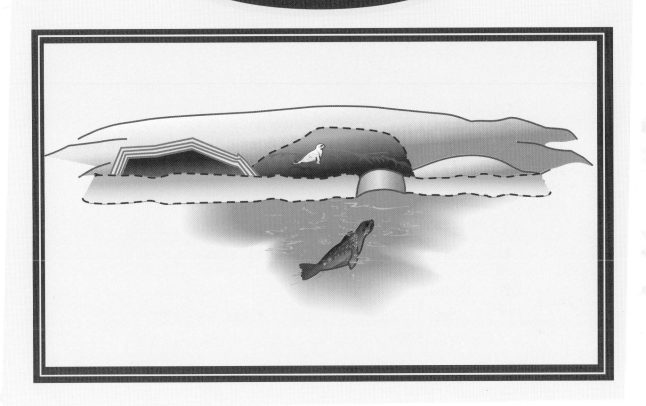

What Is So Special About Baikal Seals?

Baikal *(by KAHL)* seals are special because they live in a freshwater lake instead of a saltwater sea. That lake is Lake Baikal, the deepest lake in the world. Lake Baikal, which is in Siberia, is covered with ice much of the time. In winter, the ice may be 3 feet (90 centimeters) thick!

Baikal seals feed mostly on fish in the lake. Each seal usually makes its own breathing hole in the ice. Each seal also makes its own haul-out hole. Occasionally, two or more seals may share the holes they make.

In summer, the ice breaks up. Baikal seals haul out on rocks along the shore. Baikal seals are among the few pinnipeds that sometimes give birth to twins. Newborn Baikal pups have fur that is white and woolly. As they grow up, their coats become much darker.

Baikal seal

When Is a Harp Seal a Whitecoat?

Newborn harp seals have silky, white coats. In fact, the pups are called whitecoats. Although fur seals are born on land in rookeries, harp seals are born on pack ice. These floating blocks of ice can be crowded and noisy.

Mothers stay with their pups for about 10 to 12 days. Then they go off to find food for themselves. They do not return to their pups.

After about two weeks, a pup's white coat turns gray. The pup is also ready to swim now. It goes off to find food for itself. At first, the pup does not go very far. But as it gets older, it joins other young harp seals.

Harp seal with pup

Why Is It Easy to Recognize a Walrus?

It's easy to recognize a walrus. Walruses are the only pinnipeds with tusks. They are big animals, next in size to elephant seals.

Walruses live in the cold waters of the Arctic, North Atlantic, and North Pacific oceans. Their skin is very thick, with many folds and wrinkles. Walruses do not have much hair. But they do have a lot of blubber, which keeps them warm.

Like fur seals, walruses can walk on all fours. The rough bottoms of their flippers keep them from slipping.

If a walrus looks as if it is just standing in the water, it may be sleeping. Pouches on the sides of the male walrus's neck fill with air. The air in the pouches helps keep the animal afloat. Female walruses do not have these pouches.

Walruses

How Does a Walrus Use Its Tusks?

Most animals use their teeth for eating. You might think that walruses would use their tusks to dig for food. But guess what. They don't!

A walrus uses its tusks to defend itself from enemies, such as polar bears. Sometimes, it also uses its tusks as hooks to haul out onto the ice. At other times, a walrus uses the tusks as anchors to hold onto the ice while floating in the water.

Tusks are also good for picking at ice and making breathing holes. Tusks can even help rescue a pup that has gotten stuck in an ice crack. For example, a mother might use her tusks to destroy a block of ice to free her stuck pup.

Most walruses have two tusks. These tusks are really the animal's upper canine teeth. But what long teeth they are! The tusks may grow to be 39 inches (99 centimeters) long. Tusks continue to grow all during the animal's life.

Walrus

What—and How— Do Walruses Eat?

Walruses like to be near their favorite food—clams. A walrus will swim along in shallow water, and when it finds a clam, it will use its tongue to form a vacuum. It sucks the clam into its mouth. Then it sucks out the meat and spits out the shell. For a walrus, it takes a few thousand clams to make one good meal!

Walruses also eat fish, worms, snails, crabs, and shrimp. Sometimes, walruses eat birds. They may even eat other seals and young whales. Although walruses find most of their food in shallow waters, they can dive for food, if they need to.

Young walrus

What Is Life Like in a Walrus Herd?

Walruses like to be with other walruses. They spend most of their time in herds. A herd of walruses may drift along together on an ice floe. Or, the whole herd may haul out onto land.

There may be thousands of walruses in a herd. The animals pile up, one on top of the other. Fights sometimes break out. A large bull walrus may want a better spot in the crowd. He'll throw back his head and point his tusks at a smaller animal. The message is clear: "Move over, or else!"

Among the herd are many walrus pups, called calves. Walrus mothers care for their calves for about two years. To protect her calf in the crowd, the mother holds it under her body and between her two front flippers.

Walrus herd

Is There a Bad Guy in the Bunch?

Yes, there is, and it is the leopard seal. Leopard seals resemble leopards in the spotted pattern of their fur and the way they attack their prey.

Leopard seals have a bad reputation—partly because of their appearance and partly because of their actions. A leopard seal looks fierce. It has a big head, a wide mouth, and long canine teeth. And its long, thin body makes the leopard seal look more like a lizard than a seal. Underwater, it makes long, deep, droning sounds. It's not an animal you'd want to meet!

Leopard seals live near the South Pole. They often live alone—in the water or on pack ice. Like other pinnipeds, leopard seals swim and dive for their food. But leopard seals are also able to leap out of the water and onto ice to capture and kill their prey. They have been seen killing sea birds and penguins—as well as other seals.

Leopard seal

53

How Do Pinnipeds Measure Up?

Pinnipeds are all big animals, but they do vary in size. In length, pinnipeds can vary from about 4 feet (1.2 meters) to about 16 feet (4.9 meters). They may weigh from about 140 pounds (65 kilograms) to about 8,800 pounds (4,000 kilograms).

Ringed seals, which look plump, are among the smallest pinnipeds. Cape (South African) fur seals are longer and heavier than kinds of fur seals that live in other places. They are the largest fur seals.

Elephant seals are the only pinnipeds that are longer and heavier than walruses. Elephant seals are the largest pinnipeds in the world. Southern elephant seals live near Antarctica. They are even larger than Northern elephant seals, which live near the southwestern United States and Mexico.

The chart shows how ringed seals, fur seals, walruses, and elephant seals size up. The sizes you see are averages for adult males. Adult male pinnipeds are usually larger than adult females.

54

Comparison Chart

Length in Feet
Weight in Pounds

Ringed seal

Cape (South African) fur seal

Walrus

Southern elephant seal

4 feet
143 pounds

7 feet
660 pounds

12 feet
3,000 pounds

16 feet
8,800 pounds

How Do Scientists Study Pinnipeds?

When a pinniped is hauled out, it's easy for scientists to observe its behavior. Scientists have watched pinnipeds haul out to rest and warm themselves. Scientists have watched pinnipeds give birth, care for their young, and molt. Scientists have even observed pinnipeds escaping from enemies.

But there is still a lot that scientists do not know about pinnipeds. It's hard to observe these animals when they are not hauled out. So we do not know much about their habits at sea.

Scientists have, however, found ways to track these animals. They have attached time-depth recorders and radio transmitters to some seals. These tell where the seals swim and how deep they dive. By using these devices with satellites, scientists hope to learn even more about pinnipeds at sea.

Seal with tracking device

Do Hawaiian Monk Seals Feel Warm?

Hawaiian monk seals live in Hawaii, where the climate is warm all year. They have the same amount of blubber as seals in cooler climates. So they must try harder to cool off when they haul out.

The seals cool off by lying quietly in the shade or wet sand all day. They may lie on their dark backs with their light-colored bellies up. Their hearts beat slowly and they breathe slowly. They dive for food at night.

Hawaiian monk seals do not migrate to cooler places, but they may travel far to find food.

Long ago, there were many monk seals in Hawaii. But today, few are left.

Hawaiian monk seal

Are Pinnipeds in Danger?

In the past, some pinnipeds were in danger. They were overhunted for fur, as this baby harp seal was. Pinnipeds were also hunted for food, oil, and skins.

Hawaiian and Mediterranean monk seals are still endangered. So are Northern sea lions (sometimes called Steller's sea lions). The Caribbean monk seal has even become extinct.

As people became aware of losing whole species, they passed laws to protect the animals. People concerned with the environment also reduced the demand for seal fur. This further helped protect fur seals.

Seals have few natural enemies. They may fall prey to whales and sharks in the water. On ice, polar bears are their biggest threat. But people are the most dangerous enemy for pinnipeds. People take away their food supply. People make the waters pinnipeds swim in dangerous. So it is up to people to protect the animals.

Harp seal pup

Pinniped Fun Facts

→ Some pinnipeds eat stones and pebbles, but no one knows why.

→ On land, fur seals fan themselves with their flippers to cool off.

→ In the water, a walrus pup drinks its mother's milk while it hangs upside-down.

→ Pinnipeds eat little or nothing while they are molting.

→ The hairs in a fur seal's coat are so fine that more than 300,000 are found in a square inch (6.5 square centimeters) of skin.

→ Fur seal pups near the Antarctic often chase each other, playing a kind of tag game.

→ Walruses can eat about 6,000 clams at one feeding.

→ Pinnipeds can hear high-pitched sounds in water that people cannot hear.

Glossary

blubber Thick fat under the skin of sea animals that keeps them warm and stores energy.

breed To give birth.

endangered In danger of dying out.

energy The strength to do things.

extinct No longer existing.

fast ice Thick ice that covers water and is frozen securely to rocks along the shore.

flipper A wide, webbed foot that an animal uses for swimming.

habitat The place where an animal lives.

harbor A protected area of deep water near land.

haul out To come out of water onto land or ice.

herd A big group of animals of one kind that stay together.

mammal A warm-blooded animal with a spine and fur or hair. The young feed on the mother's milk.

migrate To move from one place to another at a special time to find food, to have young, or to adjust to changes in the weather.

molt To lose fur, skin, or another body covering before getting a new one.

oxygen A gas that is part of the air that people and animals need to breathe.

pack ice Thick ice that floats over water. It is often squeezed or piled into different formations.

pinniped In biology, the group of seals, sea lions, and walruses. Each member has four flippers and lives in water and on land.

rookery A place on land where some pinnipeds, such as fur seals, are born.

rotate To turn around.

species A group of the same kind of animals.

tusk A long, pointed tooth that comes out of the mouth of some animals.

Index

(**Boldface** indicates a photo, map or illustration.)

Picture Acknowledgments: Front & Back Covers: © Yva Momatiuk & John Eastcott, Photo Researchers; © Wolfgang Bayer, Bruce Coleman Inc.; © Gregory Ochoki, Photo Researchers; © W. Perry Conway, Tom Stack & Associates; © B. & C. Alexander, Innerspace Visions

© B. & C. Alexander, Innerspace Visions 37; © Erwin & Peggy Bauer, Bruce Coleman Inc. 49; © Erwin & Peggy Bauer, Bruce Coleman Collection 59; © Wolfgang Bayer, Bruce Coleman Inc. 4, 45, 47; © Tom Brakefield, Bruce Coleman Inc. 43; © W. Perry Conway, Tom Stack & Associates 61; © Tui DeRoy, Bruce Coleman Inc. 13; © Francisco J. Erize, Bruce Coleman Collection 53; © Michael P. Fogden, Bruce Coleman Collection 15; © Jeff Foott, Tom Stack & Associates 51; © Gilbert S. Grant, Photo Researchers 57; © Leonard Lee Rue, Bruce Coleman Inc. 21; © Joe McDonald, Animals Animals 5, 23; © Yva Momatiuk & John Eastcott, Photo Researchers 3,11; © Flip Nicklin, Minden Pictures 27, 31; © Richard T. Nowitz, Photo Researchers 25; © Gregory Ochocki, Photo Researchers 33; © Carleton Ray, Photo Researchers 35; © Hans Reinhard, Bruce Coleman Inc. 7,19; © Patricio Robles Gil, Bruce Coleman Inc. 5, 29; © L. Veisman, Bruce Coleman Inc. 41.

Illustrations: WORLD BOOK illustration by Michael DiGiorgio 16, 17, 39, 55; WORLD BOOK illustration by Karen Donica 9, 62